What Do You Know About
Life Cycles?

PowerKiDS
press.
New York

Suzanne Slade

With love to Worth and Beverly Slade, my wonderful in-laws

Published in 2008 by The Rosen Publishing Group, Inc.
29 East 21st Street, New York, NY 10010

First Edition

Editor: Amelie von Zumbusch
Book Design: Kate Laczynski
Photo Researcher: Jessica Gerweck

Photo Credits: Cover, pp. 1, 5–11, 13, 17–18, 20, 22 Shutterstock.com; p. 12 © www.istockphoto.com/Sven Peter; p. 14 © www.istockphoto.com/Nicky Gordon; p. 15 © www.istockphoto.com/Craig DeBourbon SSG C.A. DeBourbon;   p. 16 © www.istockphoto.com/David Hamman; p. 19 © animalsanimals; p. 21 © www.istockphoto.com/Shane White.

Library of Congress Cataloging-in-Publication Data

Slade, Suzanne.
  What do you know about life cycles? / Suzanne Slade. – 1st ed.
      p. cm. — (20 questions : Science)
  Includes index.
  ISBN 978-1-4042-4201-2 (library binding)
  1. Animal life cycles—Miscellanea—Juvenile literature. I. Title.
  QL49.S557 2008
  571.8'1—dc22
                                              2007034111

Manufactured in the United States of America

Contents

Life Cycles

A brand-new life begins when an animal is born or **hatches** from an egg. This new animal grows and changes over time. As animals get older, they grow stronger and begin to look different. When an animal has finished growing, it becomes an adult. The steps of change in an animal's life are called that animal's life cycle.

In time, adult animals produce their own young. Most birds, fish, and **insects** lay eggs when they are ready to start a family. Nearly all **mammals** give birth to live young. Under the watchful eyes of their proud parents, these new baby animals begin their own life cycle.

As most mammals do, deer give birth to live young. Baby deer are called fawns. While most kinds of adult deer have a plain brown coat, fawns generally have a spotted coat.

1. How long is an insect's life cycle?

The life cycle of most insects lasts only a few weeks, although some have shorter or longer lives. An insect called a cicada has the longest life cycle of all bugs. Some kinds of cicadas live for 17 years!

This bug is a 17-year periodical cicada. These insects live in the northeastern United States. They spend most of their life underground but come aboveground to have babies and die when they are 17 years old.

2. Do any insects have live young?

A few types of insects, such as aphids and small flies called tsetse flies, are able to give birth to live young.

As all beetles do, labybugs go through complete metamorphosis.

3. How do insects change as they get older?

Most insects start out as an egg and change into an adult over time. This change is called **metamorphosis**. Insects go through either incomplete metamorphosis or complete metamorphosis.

Aphids give birth to live young throughout the spring and summer. When the cold winter comes around, aphids lay eggs. The following spring, baby aphids will come out of these eggs.

4. What happens during incomplete metamorphosis?

There are three steps in incomplete metamorphosis. The insect starts out as an egg. Most mother insects do not stay around to watch over the eggs they have laid. The **nymphs**, or young insects, that hatch from an egg must look after themselves. These nymphs look much like adult insects. As a nymph eats and grows, its body becomes too large for its hard skin. The

This adult dragonfly is shedding the skin that it lived in as a nymph. Most dragonfly nymphs shed their shell more than 10 times before becoming adults!

Dragonflies are one of many insects that go through incomplete metamorphosis. Dragonflies have two pairs of wings and almost always live near water. There are several thousand kinds of dragonflies.

nymph **sheds** its tight outside skin several times before becoming an adult. Many kinds of insects, such as locusts, dragonflies, and grasshoppers, go through incomplete metamorphosis.

5. How is complete metamorphosis different from incomplete metamorphosis?

Complete metamorphosis has four steps instead of three steps. The young insect that hatches from an egg in complete metamorphosis is called a larva. A larva does not look like an adult insect at all. In time, the larva changes into a **pupa**. In the last step of complete metamorphosis, this pupa turns into an adult.

A caterpillar is the larva of a moth or a butterfly. This caterpillar will grow up to be a monarch butterfly. Monarch caterpillars eat milkweed leaves.

6. How does a larva become a pupa?

After eating a large amount of food, a larva covers itself with a hard shell and becomes a pupa. The pupa moves very little and does not eat.

Monarch caterpillars turn into a green pupa, called a chrysalis. They remain a chrysalis for about 10 to 14 days.

7. What insects go through complete metamorphosis?

Some of the insects that go through complete metamorphosis are bees, butterflies, moths, and beetles.

Adult monarch butterflies, like the butterfly to the right, have orange and black wings with white spots. Adult monarchs drink from many different flowers.

8. What other animals change a lot during their life cycle?

Animals called **amphibians** are born in water. They live in the water at first, but later, their body changes so they can live on land, too.

Most amphibians lay their eggs in the water. These eggs came from a frog, one of the best-known amphibians.

9. How do amphibians live both in the water and on land?

Amphibians lay jellylike eggs in water. A larva that looks like a small fish hatches from each egg. The larva has **gills** and a tail. During metamorphosis, the larva changes into an adult with **lungs** for breathing air. The larva also grows legs for walking on land.

10. What animals are amphibians?

Frogs, toads, and salamanders are all amphibians.

11. What is a
11. What is a frog's life cycle like?

Frog eggs hatch into fishlike larva, called tadpoles, in about two weeks. Tadpoles grow legs and lungs in the following three months and later become adults.

While most frogs spend some of their time in the water, they can also easily move around on land.

Like many animals, birds begin their life inside eggs. These eggs must be kept warm so the **embryos**, or chicks growing inside, can grow. When a chick is ready to hatch, it uses a sharp tooth on the end of its **beak**, called an egg tooth, to break open the egg. It takes a chick several hours to hatch. The newborn chick then

This tiny chick is breaking out of its egg. Chicken eggs generally hatch after about three weeks.

Some newly hatched birds have no feathers. These baby birds' eyes are not yet open, so they cannot see. They totally depend on their parents for food and safety.

rests and lets its wet feathers dry. The chick's parents bring it food until the chick is old enough to leave the nest. This young bird will grow into an adult and have its own chicks someday.

13. What is inside a bird egg?

At the center of a bird egg is a round, thick **yolk**. The yolk goes around the embryo. It also supplies food for the embryo. Around the yolk is the egg white, or albumen. The albumen keeps the yolk and embryo safe.

Bird eggs have a hard shell that keeps the embryo inside safe. Bird eggs come in many colors. Some eggs are a solid color, while other eggs have spots or other markings.

Most birds build nests for their eggs. The mother bird generally sits on the nest to keep the eggs warm. In cold Antarctica, however, father emperor penguins have the job of egg warmer. A father penguin places an egg between his feet and carefully lowers his feathered front over the egg. He warms the egg until it hatches, about 65 days later.

Some kinds of birds share the warming of eggs between both parents. For example, male and female mute swans take turns warming eggs and caring for babies.

15. How are new fish born?

A new fish shakes and wiggles to break open the soft cover of its egg.

16. Is a fish egg like other animal eggs?

As other eggs do, a fish egg has a yolk in its center. However, fish eggs do not have a hard shell, as the eggs of birds, snakes, and turtles do.

Some kinds of fish produce millions of eggs and let them drift in the water. Other kinds of fish lay their eggs on a rock or plant. These eggs are often a bit sticky, which lets them fix onto the rock or plant.

17. How does a baby fish grow into an adult?

When a fish first hatches from its egg, it is called a fry. Fry must look out for themselves. Most have a small sac of yolk on their stomach. They use this food at first. Later, fish become big enough to get their own food. The fish continue to eat and grow until they become adults.

You can see the yellow yolk sac on this arawona fry. The young fish will depend on the yolk sac for food until it is big enough to find its own food. Then, it will eat many foods, such as bugs, small frogs, and other fish.

18. Do any water animals have live young?

A few fish, such as guppies and certain sharks, have live young. Whales, dolphins, sea otters, and other water mammals all give birth to live babies.

As all water mammals do, gray seals give birth to live young. Baby gray seals are called pups. Like all gray seal pups, this baby's coat will become spotted and less furry over time.

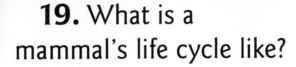
Nearly all mammals form inside their mother and are born live. Mammals care for their helpless babies and keep them safe. For example, bats keep their babies in warm, dry caves where they are safe from enemies.

New mammal babies drink the milk their mother produces. Later, young mammals eat solid food. At first, most parents bring this food to their young. Over time, young mammals grow bigger and stronger. They learn to find their own food and become adults.

Monotremes, like this echidna, are the only mammals that lay eggs. Like other mammals, baby monotremes drink their mother's milk and grow fur.

20. How long do mammals live?

Some mammals have a short life cycle, while others have a longer one. Large mammals often live longer than small mammals. For example, tiny cotton mice live only four to five months in the wild. However, giant elephants live for 55 to 70 years.

People are mammals, too. Today, many people live into their 90s. Eating good food and getting plenty of exercise can help you live a long and healthy life. As you move through your life cycle, you will discover many new and interesting things.

As many large mammals do, camels can live a long time. They often live for as long as 50 years.

Glossary

amphibians (am-FIH-bee-unz) Animals that spend the first part of their life in water and the rest on land.

beak (BEEK) The hard mouth of a bird or a turtle.

embryos (EM-bree-ohz) Plants or animals as they grow before birth.

gills (GILZ) Body parts that animals, such as fish, use for breathing.

hatches (HACH-ez) Comes out of an egg.

insects (IN-sekts) Small animals that often have six legs and wings.

lungs (LUNGZ) The parts of an air-breathing animal that take in air and supply a gas called oxygen to the blood.

mammals (MA-mulz) Warm-blooded animals that have a backbone and hair, breathe air, and feed milk to their young.

metamorphosis (meh-tuh-MOR-fuh-sus) A complete change in form.

nymphs (NIMFS) Young insects that have not yet grown into adults.

pupa (PYOO-puh) The second stage of life for an insect, in which it changes from a larva to an adult.

sheds (SHEDZ) Gets rid of an outside covering, like skin.

yolk (YOHK) The part of an egg that feeds the growing baby animal.

Index

A
adult(s), 4, 7, 9–10, 12–13, 15, 19, 21
amphibians, 12–13

B
beak, 14

C
change, 4, 7

E
egg(s), 4, 7–8, 10, 12–14, 16–19
embryo(s), 14, 16

I
insect(s), 4, 6–11

L
lungs, 12–13

M
mammals, 4, 20–22
metamorphosis, 7–12

N
nymph(s), 8–9

Y
young, 4, 6, 20–21

Web Sites

Due to the changing nature of Internet links, PowerKids Press has developed an online list of Web sites related to the subject of this book. This site is updated regularly. Please use this link to access the list:
www.powerkidslinks.com/20sci/cycle/